The Science of Structures

Janice Parker

Gareth Stevens Publishing
A WORLD ALMANAC EDUCATION GROUP COMPANY

Please visit our web site at: www.garethstevens.com
For a free color catalog describing Gareth Stevens' list of high-quality books and
multimedia programs, call 1-800-542-2595 (USA) or 1-800-461-9120 (Canada).
Gareth Stevens Publishing's Fax: (414) 332-3567.

Library of Congress Cataloging-in-Publication Data

Parker, Janice.
   The science of structures / by Janice Parker.
     p. cm. – (Living science)
   Includes index.
   ISBN 0-8368-2792-9 (lib. bdg.)
   1. Structural engineering–Juvenile literature. [1. Structural engineering. 2. Building.]
   I. Title. II. Living science (Milwaukee, Wis.)
   TA634 .P365   2001
   624.1–dc21   00-046400

This edition first published in 2001 by
Gareth Stevens Publishing
A World Almanac Education Group Company
330 West Olive Street, Suite 100
Milwaukee, WI 53212 USA

Copyright © 2001 by WEIGL EDUCATIONAL PUBLISHERS LIMITED
All rights reserved. No part of the material protected by this copyright may be reproduced or utilized
in any form or by any means, electronic or mechanical, including photocopying, recording, or by
any information storage and retrieval system, without permission in writing from the copyright
owner. Requests for permission to make copies of any part of the work should be mailed to:
Copyright Permissions, Weigl Educational Publishers, 6325 – 10th Street S.E., Calgary, Alberta,
Canada, T2H 2Z9.

Project Co-ordinator: Jared Keen
Series Editor: Celeste Peters
Copy Editor: Heather Kissock
Design: Warren Clark
Cover Design: Terry Paulhus
Layout: Lucinda Cage
Gareth Stevens Editor: Jean B. Black

Every reasonable effort has been made to trace ownership and to obtain permission to reprint
copyright material. The publishers would be pleased to have any errors or omissions brought
to their attention so that they may be corrected in subsequent printings.

Photograph Credits:
Craig Bournes, Lovat Inc.: pages 9 top left, 20; Corbis: pages 10, 11, 12 bottom right, 13 top left,
13 center, 13 bottom right, 21 left; Corel Corporation: pages 4, 5, 6, 7, 9 bottom left, 14, 16, 17, 18,
21 right, 22, 23, 24, 25, 26, 27 top left, 27 bottom left; Enermodal Engineering: page 31; Eyewire:
cover, pages 12 top right, 12 bottom left, 13 top right; Robert and Linda Mitchell: page 27 center;
William T. Peters: page 28 right; PhotoDisc: page 15; Monique de St. Croix: page 8 right; Visuals
Unlimited: pages 8 top (Pat Anderson), 8 bottom left (Glenn M. Oliver), 9 top right (A. J. Copley),
28 left (Marc Epstein), 29 right (Leonard Lee Rue Hi), 30 (A. J. Copley).

Printed in the United States of America

1 2 3 4 5 6 7 8 9 05 04 03 02 01

Contents

What Do You Know about Structures? 4

Building Materials 6

Really Big Tools 8

The Grand Design 10

The Planning Stage 12

The Building Stage 14

Parts of Structures 16

Bridges to Cross 18

Tunnels for Travel 20

Skyscrapers and Towers 22

Ancient Builders 24

The Tallest of All? 26

Structures in Nature 28

Green Structures 30

Glossary 32

Index 32

Web Sites 32

What Do You Know about Structures?

Look up. Is there a ceiling over your head? If there is, it is probably held up by walls that stand on a floor. You are inside a building of some kind. A building is a type of structure. Houses, churches, bridges, tunnels, stores, and schools are all structures. Most structures are built to be sturdy and to keep out the forces of nature, such as wind, snow, and rain.

Buildings are one type of structure.

Structures come in many shapes and sizes. Some structures provide shelter and safety. Others help us get from place to place.

Structures are made from many different materials. Some are made of clay, straw, or other materials found in nature. Others are built with **manufactured materials**, such as steel, glass, or concrete.

People have been building structures for many centuries. Structures built by people are **artificial**. All other structures are natural.

Some houses are made from materials found in nature.

Concrete and bricks are manufactured materials. They are used to make strong structures in which to live and work.

Activity

Start a Structures Scrapbook

Find pictures of structures. Magazines and newspapers are good places to look. Cut out the pictures, but be sure to ask for permission first. Paste the pictures you cut out onto the pages of a scrapbook. Label each page with the type of structure pictured.

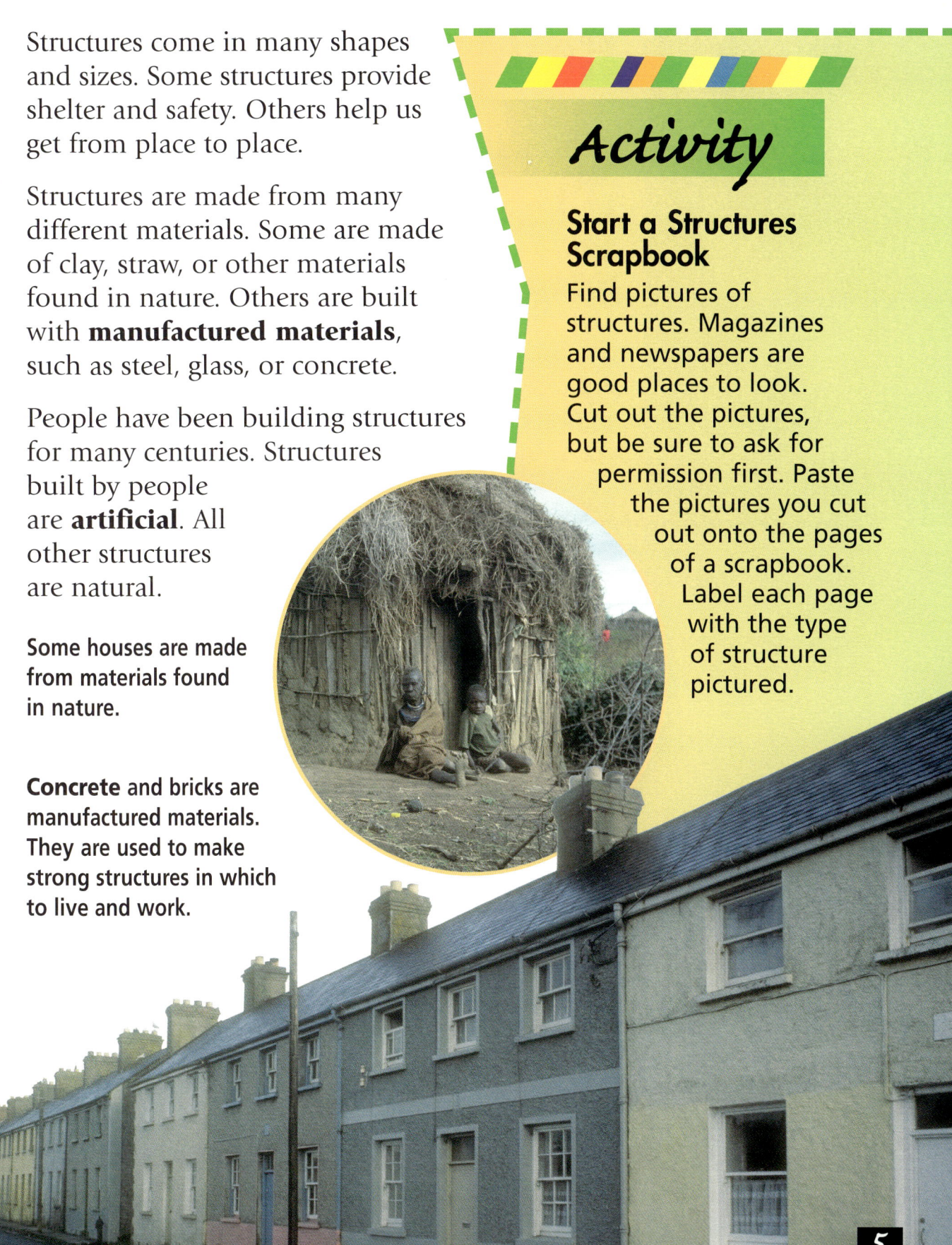

5

Building Materials

Structures can be made from many types of materials. Each material has special features that make it useful for building. Some materials are better than others for building certain kinds of structures. For example, glass is the best material for building an aquarium, which is a place where fish live. Building a house out of glass, however, is not a good idea. A glass house would be too warm inside.

Types of Building Materials

Stone	Clay	Wood

• found in nature, most often limestone • strong and long lasting • can be used whole or cut into blocks	• soil that has fine particles • soft and sticky when wet, hard when dry • can be made into bricks	• comes from trees • easy to use • can be held together with nails

Examples

pyramids of Egypt, castles, marble buildings	brick houses, chimneys, fireplaces	log cabins, beaver lodges, bird nests

Puzzler

From which material is each of the following structures built?
- a rocking chair
- a termite mound
- a tent
- an igloo
- a bird's nest

Answer:
- a rocking chair: wood
- a termite mound: soil
- a tent: canvas or nylon cloth
- an igloo: snow
- a bird's nest: twigs

Concrete

- like stone, but made by people
- strong when dry
- can be poured into any shape when wet

office buildings, house foundations

Glass

- made from melted sand
- light shines through it
- can be made in many colors

greenhouses, skylights, windows

Steel

- used in most large buildings today
- very strong
- can be melted and molded into shapes

skyscraper frames, bridges, the Eiffel Tower

7

Really Big Tools

Humans have used tools for many centuries. Tools help build bigger and better structures. Building today's huge structures requires really big tools and specially trained people to use them.

Bulldozers
clear away soil and rubble from construction sites. They are used to flatten a large piece of land on which a structure will be built.

Backhoe Loaders
dig large holes in the ground. These holes can be filled with concrete to make **foundations** for houses and other buildings.

Excavators
are like backhoe loaders, but they are much larger. They are used to dig large, deep foundations for structures such as skyscrapers or underground parking lots.

Tunnel-boring Machines (TBMs) are giant drills.
They dig tunnels underground by twisting and turning as they grind their way through soil and rock.

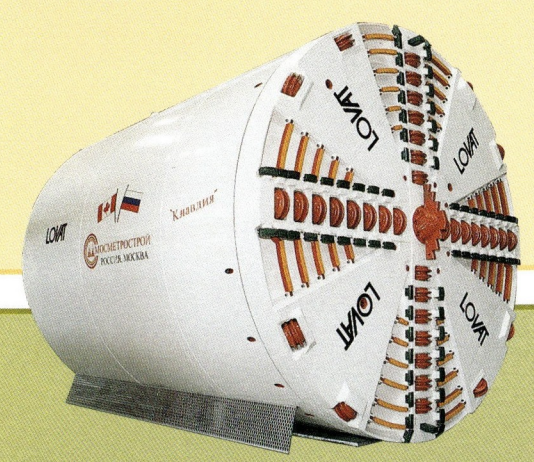

Pile Drivers
are huge hammers that pound big upright beams, called piles, deep into the ground.
Piles support heavy structures. A pile driver has a heavy weight that is lifted into the air, then dropped onto a pile. The weight drives the pile into the ground.

Cranes
are tall lifting machines. They raise heavy building supplies to the tops of high structures.

Activity

Watch a Construction Site
The next time you see a structure being built, stop and watch — but stand back! Construction sites are dangerous places. Look at the tools being used. Do you see any of the machines shown on these pages? Do you see any other machines or tools? Try to figure out the purpose of each machine or tool you see.

The Grand Design

Would you like to design structures? People who do this kind of work are called **architects**. The job of an architect combines both art and science. Architects design structures that are interesting to look at. They also know how to make structures safe.

Before a structure can be built, many questions must be answered. Architects help answer these questions:

- Where should the structure be built?
- How much money will the structure cost?
- How can the structure be made to look attractive?
- What kind of building materials should be used?
- What is the best way to build a safe structure that will last a long time?

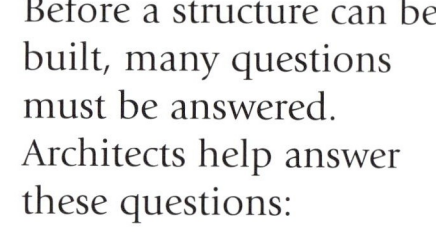

Architects are hired by people who want to build structures.

Architects first make drawings or models of a structure to show what the structure will look like when it has been built. Today, computer models are often used to show a **three-dimensional image** of a completed structure.

Architects earn university degrees. Then, after working with experienced architects, they take tests. If they pass the tests, they receive a license to work on their own.

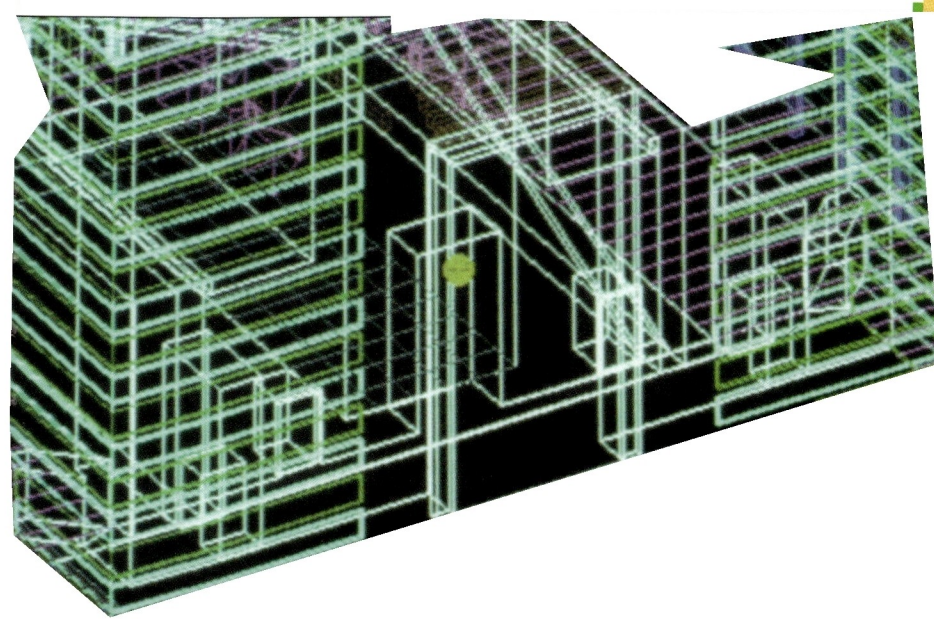

It takes patience and skill to design safe structures.

Activity

Do Your Own Research

Many jobs involve the design or construction of structures. Ask a parent or a teacher to help you find out about the following jobs:

- building contractor
- carpenter
- civil engineer
- construction foreman
- draftsperson
- interior designer
- landscape architect
- steelworker
- roofer

A computer drawing helps an architect see a structure before it is built.

The Planning Stage

A lot of work goes into building a large, complex structure. It all begins with planning. Many people with many different skills work together through the planning stage.

Clients
decide they want to have a structure built.
They have ideas about what the structure should be like. They also have an idea of how much money they want to spend to build the structure.

Architects
work with clients to design a structure.
They create drawings and models for clients to look at. They also make changes when clients ask for them.

Engineers
work with architects on large projects.
They select the best materials to use. They also choose any equipment the structure will need, such as elevators.

Draftspersons
create detailed drawings of the structure.
They need special tools and information collected from architects and engineers to complete the drawings.

Building Contractors
are in charge of getting the structure built.
They find out how much the structure will cost and how long the project will take to complete. They also hire the people needed to build the structure.

Interior Designers
decide how the inside of the structure will look.
They choose colors for the carpets and walls of a building. They might also choose some of the furniture.

Landscape Architects
make the grounds around a structure look nice.
They design fountains and green spaces. They also choose trees and other plants.

Puzzler
Imagine you want to build a bridge. Which people on these pages will not be needed for the project?

Answer: Interior designers will not be needed to build a bridge.

The Building Stage

Materials, machines, and builders all come together at the building stage. Building starts when the designing and planning stages are finished. The building stage includes all of the work that turns materials into a finished project — a structure.

A building contractor is in charge of this stage but usually hires a superintendent to help. The superintendent orders materials, gets equipment, and hires the workers. The superintendent also draws up a schedule that shows when each step of the work should be completed.

Building a structure takes time and skill.

Different tradespeople do different jobs. Welders join together pieces of metal.

For each type of work needed to complete the structure, the superintendent gets help from a person called a foreman. One foreman might be in charge of all the concrete workers. Another foreman might be in charge of all the steelworkers.

The workers, or tradespeople, who build the structure each have special skills. They might be concrete workers, welders, carpenters, or truck drivers.

Activity

Plan and Build a Structure

What you will need:
- paper
- pencil
- building materials (e.g., Popsicle™ sticks, glue, wooden blocks, cardboard, clay, Styrofoam™, and tape)

1. Decide on the kind of structure you want to build. You might want to make a small house, a big barn, a bridge, or a skyscraper.
2. Make a drawing of what the structure should look like when it is finished. Also, try to figure out how long it will take to build the structure.
3. Choose the materials you want to use.
4. Build the project. Does it look like your drawing? Did it take as long to build as you thought it would?

Carpenters work mainly with wood. They use tools such as hammers, nails, drills, and power saws.

Parts of Structures

Structures are made of many parts. Each part must be strong and must work well for a structure to be safe and sound. Every part of a structure has a special purpose. A foundation is at the bottom of a structure. It supports the structure's weight. The foundation also holds the structure together in case the ground beneath it or around it moves.

Structures are built from the ground up, starting with the foundation.

Strong materials, such as wood or metal, are joined together to create the framework of a structure. The framework holds up the structure and creates its shape. Wooden **beams** are nailed together to build the framework of most houses in North America.

Sheets are flat pieces of wood, metal, or other materials used to cover the outside of a structure. They help protect the inside of the structure from wind, snow, and rain. Roof shingles are an example of sheets.

Insulation is placed between the outside and inside surfaces of a building. Insulation keeps out heat and cold.

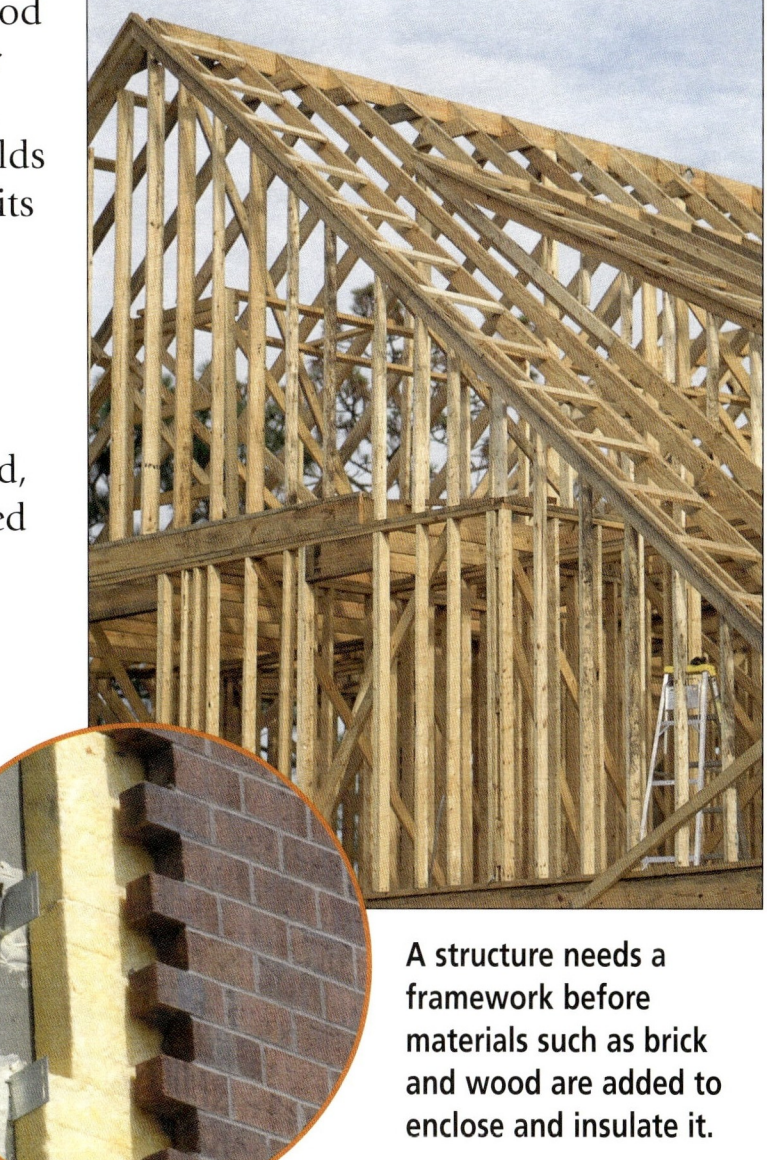

A structure needs a framework before materials such as brick and wood are added to enclose and insulate it.

Puzzler

Which part of the human body acts as a framework? Which part helps protect people from wind, snow, and rain? Which part insulates people from heat and cold?

Answer:
A skeleton is the framework of the human body. Skin protects the body, and body fat insulates it.

Bridges to Cross

Bridges are structures built across rivers, **ravines**, and bays. They help people travel quickly and safely from one place to another.

Early bridges were often fallen trees placed across small streams. Only one or two people could cross at a time. Today's bridges are much larger and stronger. Most bridges are for people and cars. Some are just for trains. Others can support people, cars, and trains — all at the same time!

A suspension bridge hangs from very strong ropes called **cables**. The cables are attached to tall posts or towers. Sometimes trees are used as posts for small bridges. Steel towers support larger suspension bridges. A suspension bridge can be wider and longer than any other kind of bridge.

The Golden Gate Bridge in California is a suspension bridge. It spans 4,200 feet (1,280 meters) over part of San Francisco Bay.

A **girder**, or beam, bridge is the most simple kind of bridge. It is a flat roadway that rests on **piers**.

Arch bridges are stronger than girder bridges. Because they can support more weight, arch bridges can be longer than girder bridges. Arch bridges can also be higher than girder bridges. They can be built high enough for tall ships to pass underneath.

girder bridge

arch bridge

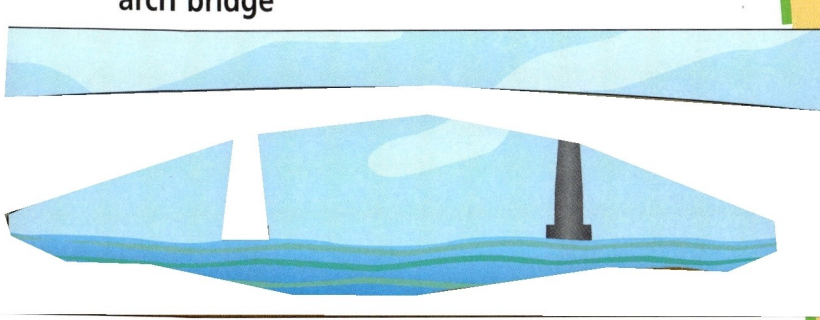

suspension bridge

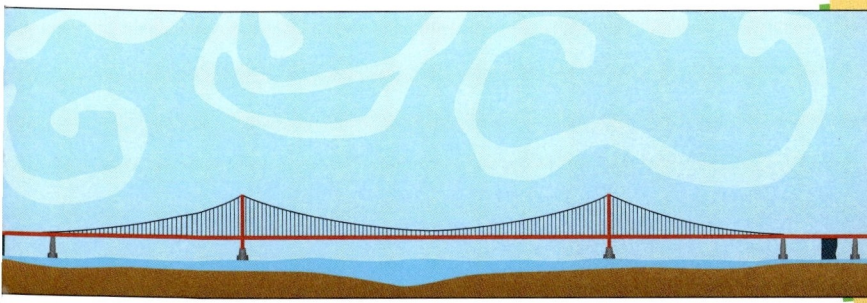

Activity

Build a Girder Bridge

What you will need:
- four thick, heavy books, all the same size
- a piece of paper
- a piece of cardboard
- a large, thin book with a hard cover, such as a picture book
- small objects to place on the bridge

1. Place two of the heavy books side by side, leaving a gap between them.
2. Lay the paper on the books so it crosses the gap.
3. Place another heavy book on top of each of the first two books.
4. Place an object on the paper bridge. Is the paper strong enough to hold the object? If so, add more objects until the bridge falls apart.
5. Now make a bridge with a piece of cardboard instead of paper. Then make one using the thin, hardcover book. How much weight can these bridges hold?

Tunnels for Travel

Tunnels are structures built underground or underwater. Some tunnels go through mountains or under cities. Others go beneath rivers and channels. Tunnels make travel through these areas faster and safer. Roads and subways go through tunnels. Some tunnels also carry water, waste, or wiring.

Building a tunnel can be dangerous. Engineers must plan carefully. Tunnel-boring machines are used to make large holes underground, but they cannot dig through very hard rock. To dig through rock, engineers drill small holes in the rock and fill the holes with explosives. The explosives blast the rock apart, creating part of a tunnel — and a big mess! All of the rubble from the explosion must be removed to clear the tunnel.

Tunnel-boring machines dig through earth so underground structures can be built.

While digging a tunnel, workers use a metal shield to keep soil and rocks from falling onto them and their machines. When the tunnel is finished, supports and a concrete ceiling keep the tunnel from collapsing.

Tunnels are often built from both ends at the same time. Engineers must make sure that the two sections meet perfectly underground.

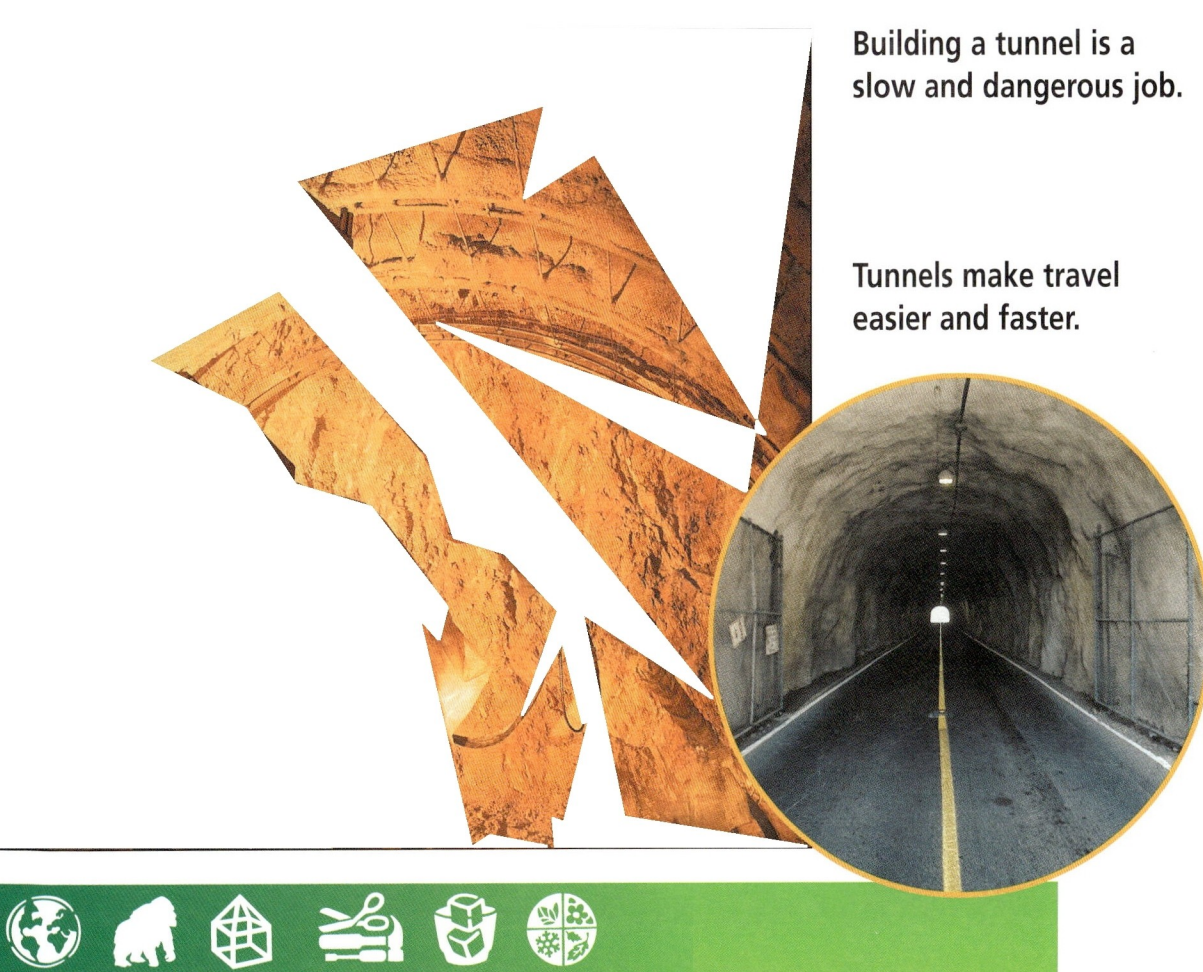

Building a tunnel is a slow and dangerous job.

Tunnels make travel easier and faster.

Puzzler

Many animals are good tunnel diggers. Can you think of any?

Answer: Moles, prairie dogs, and even some spiders are good tunnel diggers.

Skyscrapers and Towers

Some cities have little room for new buildings. The only way to build large new structures is straight up. Skyscrapers and towers are the tallest artificial structures in the world.

Skyscrapers are tall buildings with many floors stacked on top of each other. These buildings contain homes, offices, restaurants, and stores. Towers are tall structures, too, but they usually do not have offices or homes in them. Some towers have no rooms in them at all.

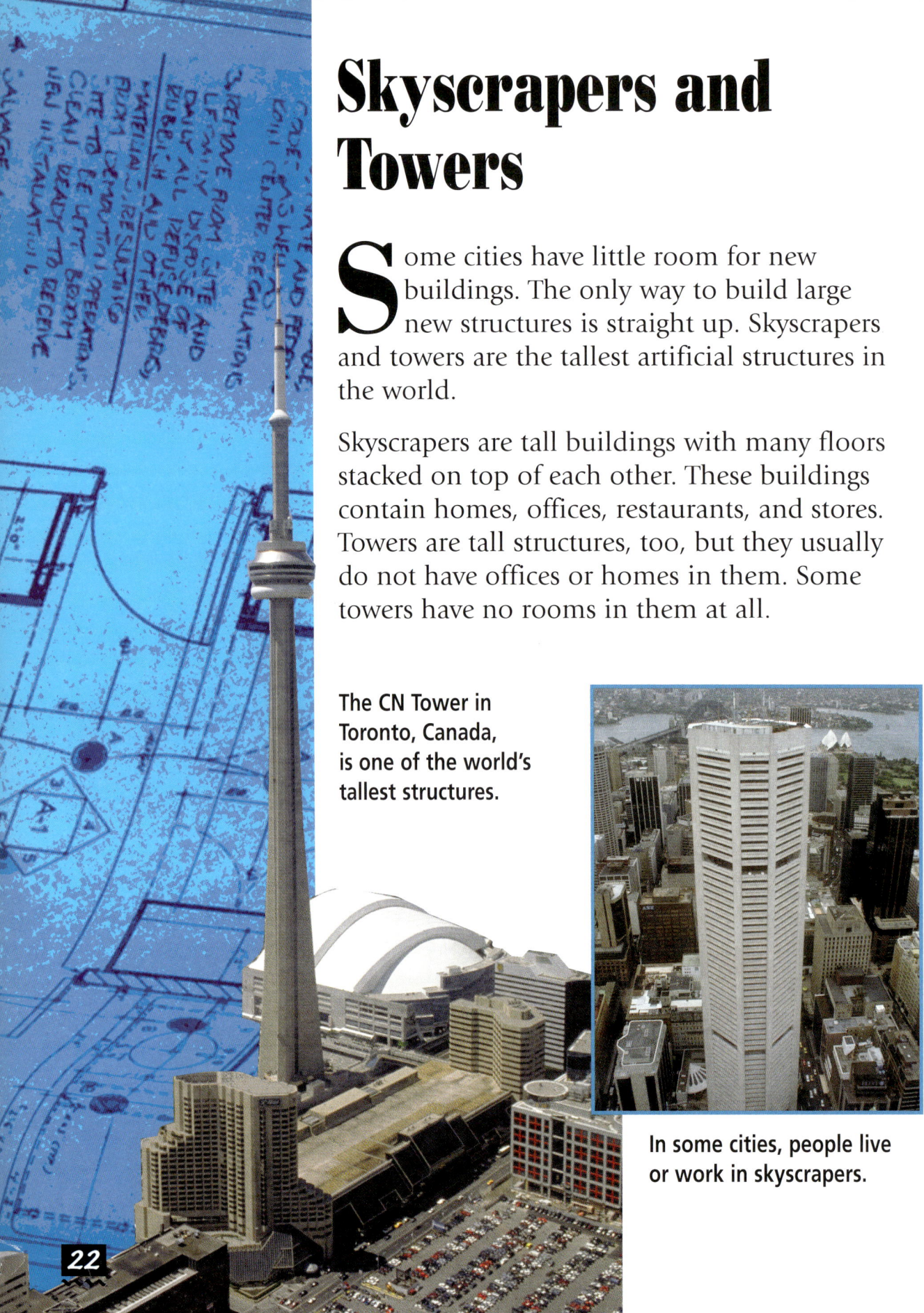

The CN Tower in Toronto, Canada, is one of the world's tallest structures.

In some cities, people live or work in skyscrapers.

Tall structures are very heavy. They contain a lot of building materials. What keeps them from breaking down under their own weight? Most skyscrapers and towers have a strong framework. It is made of steel and goes deep into the ground. Also, the top sections of skyscrapers and towers are often made of materials that weigh less than those used to build the bottom sections.

Puzzler

Towers often have equipment at the very top. Do you know what that equipment is used for?

Answer: The equipment at the top of towers usually sends and receives electronic signals from satellites and television or radio towers. Because towers are so high, the signals can come and go over the tops of other buildings.

The Eiffel Tower in Paris, France, is perhaps the most famous tower in the world.

Ancient Builders

People have been building large structures for a very long time. The pyramids in Egypt were built more than 4,500 years ago. The largest pyramid is called the Great Pyramid. It contains nearly 2.5 million limestone blocks!

It took 100,000 men twenty years to build the Great Pyramid. Workers used only simple tools and ropes. They measured and carved out each block of stone. Then they moved the stones to the building site. It took many men to pull the heavy blocks over log rollers.

The Great Pyramid was the tallest artificial structure until the Eiffel Tower was built in 1889.

How did the workers raise the stone blocks to the top of the pyramid? Many historians believe they dragged the blocks up slanting ramps. When the pyramid was finished, the ramps were removed.

The ancient Chinese built the Great Wall to protect themselves from enemies.

Over 2,000 years ago, the Chinese built a large structure called the Great Wall. This wall stretches about 1,500 miles (2,400 kilometers) across China. The Great Wall is made of **granite** blocks and rubble. It is 20 to 30 feet (6 to 9 m) high. In some places, it is two walls with a gap between them. The gap is filled with earth.

Puzzler

The pyramids of Egypt are among the seven wonders of the ancient world. Can you guess what the seven wonders of the modern world might be?

Answer:
There are many lists of the world's modern wonders. Not all of these lists are the same. Here is one list:
- Channel Tunnel, English Channel, UK/France
- CN Tower, Toronto, Canada
- Empire State Building, New York City, USA
- Golden Gate Bridge, San Francisco, USA
- Hoover Dam, Colorado River, USA
- Itaipu Dam, Parana River, South America
- Panama Canal, Panama, Central America

Can you think of other structures you would put on this list?

25

The Tallest of All?

Some of the most famous structures in the world are also the tallest — or were the tallest for a while. Every day, engineers work on new projects all over the world. They are always thinking up ways to build structures that are even taller than the tallest!

Great Pyramid (Egypt)

The Great Pyramid was built between 2660 BC and 2560 BC. It is the largest stone structure in the world. When it was built, it was 481 feet (147 m) high.

Eiffel Tower (France)

The Eiffel Tower was completed in 1889. At 984 feet (300 m) tall, it was almost twice as high as any other structure.

Empire State Building
(USA)

The Empire State Building was finished in 1931. It was the tallest building in the world for more than forty years. This structure is 1,250 feet (381 m) tall, and it has seventy-three elevators.

Sears Tower
(USA)

In 1974, the Sears Tower became the tallest building in the world. It is 1,450 feet (442 m) tall.

Activity

Find Tall Structures
Keep a list of all the tall structures you see in one week. The structures can be artificial or natural. Try to guess how tall each structure is. Compare the size of each structure to the size of your home.

Petronas Towers
(Malaysia)

Built in 1998, the Petronas Towers are now the tallest structures. They are 1,483 feet (452 m) tall.

Structures in Nature

Some animals are great architects. They build structures that help them survive in their surroundings and stay safe. Most animal homes are made of strong materials. Some contain many **chambers**.

Honeybees make their homes out of wax. They build many chambers where they lay eggs and store honey.

A termite mound can be as large as a house. Up to ten million termites can live in the chambers inside!

Orioles make nests that look like baskets. It takes an oriole four to fifteen days to build a nest.

Beavers build dams and lodges in lakes and streams. The inside of a beaver lodge has a living chamber, a food chamber, and even a bathroom!

Puzzler

Which insect makes its home out of paper?

Answer: A paper wasp chews bits of trees and grasses, which mix with the insect's **saliva** to form a paste. When the paste dries out, it is a strong paper. The wasp layers this paper to build a large nest with many chambers.

29

Green Structures

Architects face many design challenges. One is to create "green" structures, which are buildings that do as little damage to the **environment** as possible. Power for green structures often comes from **renewable** energy sources. Solar power and wind power are renewable sources of energy.

How can structures harm the environment? Forests are cut down to build structures. Some building materials release dangerous gases into the air. Energy is needed to heat and light homes and offices. Burning fuels to create some of this energy pollutes the environment.

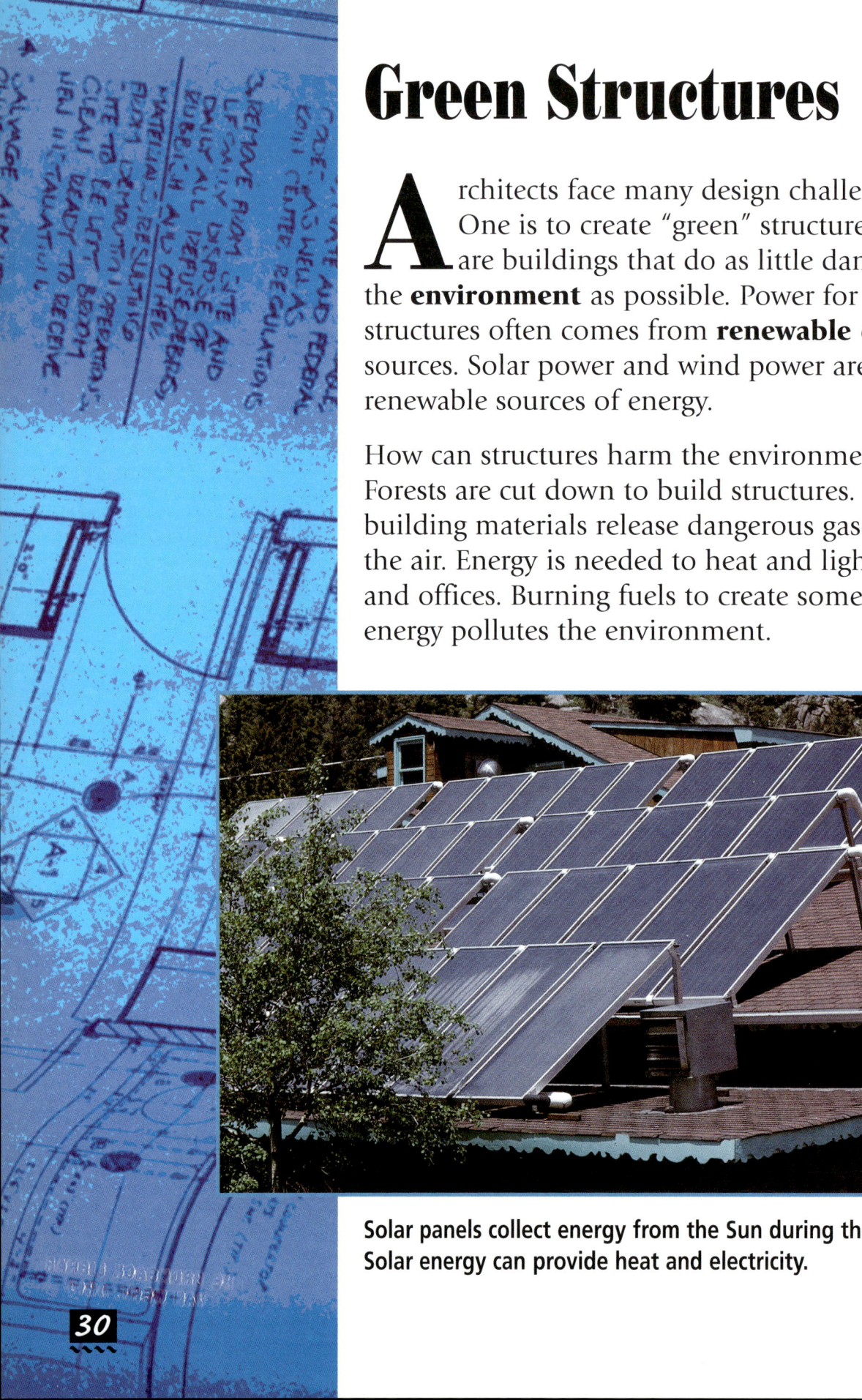

Solar panels collect energy from the Sun during the day. Solar energy can provide heat and electricity.

30

The Green on the Grand building in Kitchener, Canada, is a green structure. It uses half the energy and water that most similar buildings use.

You can help keep the environment green. Follow these simple rules to avoid wasting energy and materials.

- Turn off the lights whenever you leave a room.
- Turn off televisions, radios, and computers when you are not using them.
- Instead of turning up the heat during cold weather, put on a sweater.
- Recycle newspapers, aluminum cans, plastic, and glass.

Glossary

architects: people who design structures.
artificial: not natural; made or manufactured by people.
beams: long, slender pieces of strong materials, such as wood and metal.
cables: thick, strong steel or fiber ropes.
chambers: rooms or other enclosed spaces.
concrete: a hardened mixture of sand, gravel, cement, and water.
environment: the natural surroundings of a person, plant, or animal.
foundations: solid bases that support the weight of structures.
girder: a long steel or wooden beam that supports part of a structure.
granite: a type of rock that is very hard.
insulation: a material that heat or sound cannot easily pass through.
manufactured materials: products made by humans.
piers: pillars that support a bridge.
ravines: steep valleys.
renewable: able to be used over and over.
rubble: rough pieces of broken rock.
saliva: a watery fluid that forms inside the mouth.
three-dimensional image: a picture that shows an object's height, width, and depth.

Index

animals 21, 28, 29
architects 10, 11, 12, 13, 28, 30

backhoe loaders 8
beams 9, 17, 19
beaver lodges 6, 29
bird nests 6, 7, 29
bricks 5, 6, 17
bridges 4, 7, 13, 15, 18, 19, 25
bulldozers 8

carpenters 11, 15
clay 5, 6
concrete 5, 7, 8, 15, 21

contractors 11, 13, 14
cranes 9

draftspersons 11, 13

engineers 11, 12, 13, 20, 21, 26
environment 30, 31
excavators 8

foremen 11, 15
foundations 7, 8, 16
frameworks 7, 17, 23

glass 5, 6, 7, 31

houses 4, 5, 6, 7, 8, 15, 17, 28

insulation 17
interior designers 11, 13

metal 14, 17, 21

piers 19
pile drivers 9
pyramids 6, 24, 25, 26

ramps 25
roofers 11
rubble 8, 20, 25

sheets 17
skyscrapers 7, 8, 15, 22, 23, 27

steel 5, 7, 18, 23
steelworkers 11, 15
stone 6, 7, 24, 25, 26
superintendents 14, 15

termite mounds 7, 28
tools 8, 9, 13, 15, 24
towers 7, 18, 22, 23, 24, 25, 26, 27
tradespeople 14, 15
tunnels 4, 9, 20, 21, 25

walls 4, 25
welders 14, 15
wood 6, 7, 15, 17

Web Sites

www.archkidecture.org

www.pbs.org/wgbh/buildingbig

www.pbs.org/wgbh/nova/bridge

www.yesmag.bc.ca/focus/structures/structures.html

Some web sites stay current longer than others. For further web sites, use your search engines to locate the following topics: *architects*, *bridges*, *buildings*, *construction*, *machinery*, and *structures*.

32